ANIMAL RHYME TIME!

RHYME TIME WITH TOADS!

BY LEIGH MCCLURE

Gareth Stevens PUBLISHING

Please visit our website, www.garethstevens.com. For a free color catalog of all our high-quality books, call toll free 1-800-542-2595 or fax 1-877-542-2596.

Library of Congress Cataloging-in-Publication Data

Names: McClure, Leigh, author.
Title: Rhyme time with toads! / Leigh McClure.
Description: Buffalo, NY : Gareth Stevens, [2024] | Series: Animal rhyme time! | Includes index.
Identifiers: LCCN 2022052664 (print) | LCCN 2022052665 (ebook) | ISBN 9781538285398 (library binding) | ISBN 9781538285381 (paperback) | ISBN 9781538285404 (ebook)
Subjects: LCSH: Toads–Juvenile literature. | Rhyme–Juvenile literature.
Classification: LCC QL668.E227 M39 2024 (print) | LCC QL668.E227 (ebook) | DDC 597.8/7–dc23/eng/20221117
LC record available at https://lccn.loc.gov/2022052664
LC ebook record available at https://lccn.loc.gov/2022052665

Published in 2024 by
Gareth Stevens Publishing
2544 Clinton Street
Buffalo, NY 14224

Designer: Claire Wrazin
Editor: Caitie McAneney

Photo credits: Cover David Havel/Shutterstock.com; p. 5 Johan Larson/Shutterstock.com; p. 7 Miles Boyer/Shutterstock.com; p. 9 (top) WildlifeWorld/Shutterstock.com; p. 9 (bottom) ClarkWarren1991/Shutterstock.com; p. 11 Lutsenko_Oleksandr/Shutterstock.com; p. 13 Martin Grossman/Shutterstock.com; p. 15 Laura Ojalvo Ortega/Shutterstock.com; p. 17 (main) Amy Lutz/Shutterstock.com; p. 17 (inset) Astrid Gast/Shutterstock.com; p. 19 Nancy Bauer/Shutterstock.com; p. 21 noppasit TH/Shutterstock.com.

Printed in the United States of America

CPSIA compliance information: Batch #CS24GS: For further information contact Gareth Stevens, New York, New York at 1-800-542-2595.

CONTENTS

Boldface words appear in the glossary.

Awesome Amphibians!

What's that animal on the road?

Don't step on it! It's a toad.

It's an **amphibian**, much like a frog.

Cane toads grow as big as small dogs!

Other toads come in much smaller sizes.

Turn the page for more toad surprises!

CANE TOAD

All About Toads

Toads are often easy to find.

Earth has more than 300 kinds.

True toads use short legs to get around.

Their dry skin is often shades of brown.

Glands behind its eyes and down its back

let out **poison** when a toad is **attacked**.

AMERICAN TOAD

Toads vs. Frogs

Toads crawl, while frogs like to hop.

Toad heads have warts on top.

Frogs are slimy without masses, or lumps.

Toads have dry skin covered in bumps.

Frogs need water. They rarely stray.

Toads can drift a bit farther away.

FROG

TOAD
WART

Where in the World?

A toad isn't a **rare** animal, is it?

No, toads live in most places you visit.

American and European toads are a common sight.

South American cane toads poison animals that want to try a bite.

Toads are also found in Asia and Africa.

But you won't find a toad in Antarctica!

EUROPEAN TOAD

Moist Habitats

Moist places are often a toad's home.

Grasslands are a good place to **roam**.

Some live in backyards and beyond.

Some live in woodlands or by ponds.

Some live in wetlands and forests too.

Are there toads living near you?

Tasty Treats

Toads eat a lot for such small creatures.

Sticky tongues are a fun toad feature!

They'll eat bugs and worms in one bite.

Snails and slugs can't put up a fight.

To eat food whole, wide mouths are **ideal**.

Some toads even make mice their meal!

A Toad's Life

Toads have many sons and daughters.

They lay up to 30,000 eggs in water.

Tadpoles, or babies, **hatch** from eggs.

Baby toads have a tail instead of legs.

They swim in water a few months more.

Then they grow legs to live on shore!

TADPOLES
TOAD EGGS

Best Defenses

"Will I get warts if I find a toad to hold?"

No, that's an old story someone told.

But toads do have a great **defense**.

Their skin lets out poison if they're **tense**.

They puff up their bodies to look large.

That shows predators who's in charge!

Blending In

Would-be predators must pass this test: find toads that **blend** into the forest. Toads have colors that help them hide. Look at mud. Try to spot a toad inside! Next time you step through leaf litter, keep an eye out for this cool critter!

GLOSSARY

amphibian: An animal that spends time on land but must have babies and grow into an adult in water.

attack: To try to harm someone or something.

blend: To match with the surroundings.

defense: A way of guarding against an enemy.

gland: A body part that produces something needed for a bodily function.

hatch: To break open or come out of.

ideal: Exactly right or fitting for a person or purpose.

moist: Containing water.

poison: Something that causes illness or death.

rare: Uncommon or special.

roam: To wander.

tense: Feeling nervous.

FOR MORE INFORMATION

BOOKS

Shaffer, Jody Jensen. *Cane Toads.* Minneapolis, MN: Kaleidoscope Publishing, 2022.

Weingarten, E. T. *I See a Toad.* New York, NY: Gareth Stevens Publishing, 2019.

WEBSITES

Cane Toad
kids.nationalgeographic.com/animals/amphibians/facts/cane-toad
Learn more about the huge cane toad with National Geographic Kids.

Frogs and Toads—Explaining Their Differences
easyscienceforkids.com/all-about-frogs-and-toads/
Learn more about the differences between frogs and toads.

Publisher's note to educators and parents: Our editors have carefully reviewed these websites to ensure that they are suitable for students. Many websites change frequently, however, and we cannot guarantee that a site's future contents will continue to meet our high standards of quality and educational value. Be advised that students should be closely supervised whenever they access the internet.

INDEX